SAVE OUR PLANET

RE CLAIM VAMP PURPOSE

10 WAYS TO PRACTICE FASHION SUSTAINABILITY

INDEPENDENT AUTHORS
PUBLICATIONS

For more information contact:
Independent Authors Publications
PO Box 7062,
Roselle, NJ 07203
www.independentauthorspublications.com

Cover Design – WebprintLab
Edited – Catherine Felegi

Print ISBN: 978-1-950974-24-5
Digital ISBN: 978-1-950974-25-2

Dedication Page

Save Our Planet Reclaim Revamp Repurpose: 10 Ways to Practice Fashion Sustainability is dedicated to my children, Lucas, Tiara, and Daniel. They continue to be my sounding board for various decisions when it comes to my creativity. I love them very much and pray for their success and joy.

Table of Contents

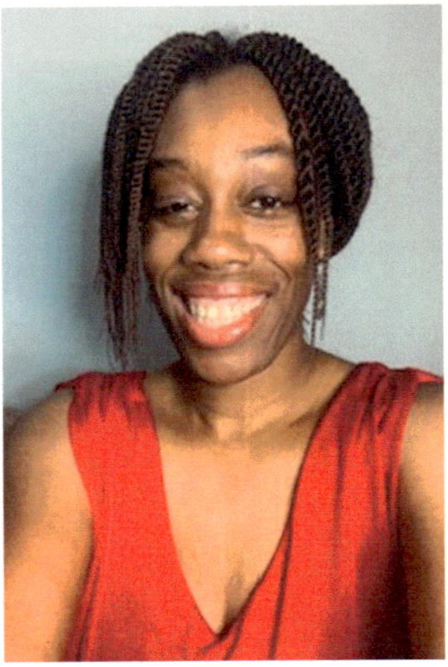

Meet DC, Queen of 3 R's (Reclaim-Revamp-Repurpose). At just seven years old, DC and her older siblings were left with family while their parents made their way to the US. DC's parents didn't have much money to spare because they were working two jobs each to save money to bring her and her two siblings to America.

DC loved to play, sing, and dance. One day, she wanted a doll to play with, but she was sadly denied because her parents didn't have the money to buy one and ship it to her. Her mom told her that she needed to wait until after she came to the US and when she was able to, she would buy DC many dolls.

Well, DC decided to take matters into her own hands. Her grandma's house had a couple of mango trees. For a treat, she climbed one of the trees and got a mango. She happily ate the fruit. Afterwards, she looked at the core of the mango and decided she would make her own doll instead. You see, the mangos in Haiti are different from other mangos because the pit has lots of strings. DC decided that she would not throw the pit away, but she would braid the "hairs" and play with it. She even named the new doll Marie. It had no hands and no feet, but lots of “hair”. She played with it for a couple of days but had to throw Marie out when she started drying out. And so that's how DC’s story began.

Throughout her life living in the US, DC would use her creativity to make beautiful items with what she had at home with a little help from the dollar stores and later her cousin, the late Kerlande Ceus, who was the owner of a flower and bridal shop. In high school, DC’s father started giving her $2 every weekday to buy snacks after school. At certain times of the year, DC would not eat any snacks so that she could buy items from the dollar stores to adorn her latest designs for birthday and holiday gifts to be given to her friends and family members. She had no idea that she was practicing upcycling and repurposing.

DC went to college and received several degrees and certificates: Fashion Design, Business Marketing, a Master’s in Special Education, Entrepreneurship, Life Coaching, and ESL. She has not lost her knack for creativity because she continues to practice repurposing and upcycling and has taught others to do the same. Indeed, DC reclaims clothing that would normally go to landfills and turns them into other items. In the future, she wants to have her own boutique where she will sell her items to the public.

Ways to join DC in her efforts for promoting sustainable fashion:

There are ten ways that you can join DC in her efforts for promoting sustainable fashion.

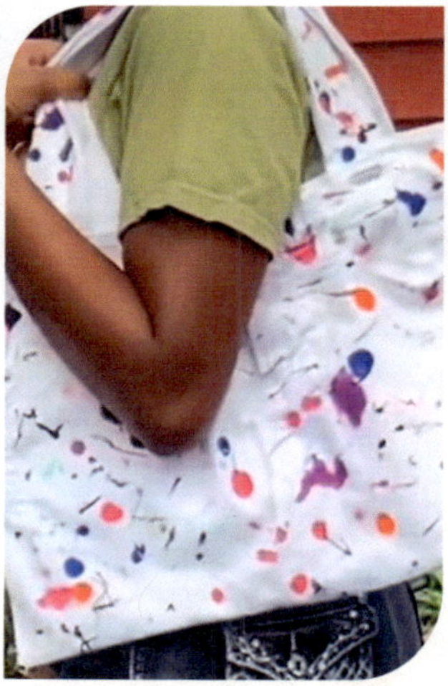

1. Shop at thrift shops

You can shop at secondhand stores, vintage shops, or online thrift stores. This helps reduce the demand for new clothing items and diverts textiles from landfills.

Most thrift stores will have things like clothes, houseware items, knick knacks, and shoes. Each item is usually tagged with a particular color and many shops will choose at least one color to be items that are on sale. One rule of thumb rings true all across every thrift store: if you see something that you like, go ahead and get it because it most likely won't be there next time you visit again. I always keep a few plastic garbage bags in my car to be used at my favorite thrift stores. It's important to tie the bag that contains your purchases tightly before you place it in your car.

Look in your neighborhood for thrift stores. You will be able to find so many treasures.

2. Rent or borrow clothing

Instead of purchasing new clothing items for special occasions, rent or borrow clothing from friends or renting services. This can reduce the number of clothing items that are only worn once and promotes the sharing economy. Below is a list of websites to rent clothes from. Each of them has their own way of conducting business with them. Also, be ready to deal with the fact that sometimes these websites shut down and the business closes. Read on their reviews to determine which one you want to do business with.

Borrow clothes from your bestie or rent clothing from online stores such as:

Fashionpass.com
Nuuly.com
Renttherunway.com
Armoire.com
Vivrelle.com
Switch.com

3. Support sustainable fashion brands

Fashion sustainability is using items that are ethically produced. Choose to support sustainable fashion brands that prioritize ethical and sustainable production practices. Research brands that use environmentally friendly materials, recycle textiles, or prioritize fair labor practices.

There are many clothing brands that are leading the way to sustainable fashion:

1. PACT
2. Patagonia
3. Tentree
4. Allbirds
5. Kotn
6. Outerknown

4. Clothing repair or upcycling

Get your clothes repaired or upcycle your clothing items instead of throwing them away. This can be done through sewing, embroidery, and patching. You can upcycle your clothes or get someone else to do it for you. Upcycle is taking an item and turning it into something new. This will extend the life of a clothing item and reduce the need for new purchases.

You can learn how to repair your clothes or you can go to your nearest dry cleaner that offers tailoring work. You can even search YouTube to find ways to upcycle your clothes. There are various DIY videos that can teach you how to upcycle without sewing being involved. In doing this, YouTube will be your best teacher. I have learned many things through YouTube and you can, too. Some upcycling requires the use of a sewing machine but many only take needle and thread.

5. Clothing swaps or donations

Have clothing swaps or donate clothing items during every season of the year. This will provide an opportunity to find new clothing items and keep textiles out of landfills. You need a team of people to help organize a clothing swap. You can conduct a clothing swap among your family or friends or your community. Donating your time at a clothing swap or just donating your clothes will make you feel good about your act of kindness toward the environment.

Hold a clothing swap event with your community this coming season.

Participate in a clothing swap with friends or organize a clothing swap event at school.

1. Plan ahead: Decide on a date and time for the clothing swap at least a few weeks in advance. This will give everyone time to clean out their closets and gather their items to swap.

2. Set ground rules: It's important to set some ground rules to ensure that the clothing swap is fair and enjoyable for everyone. For example, you may want to set a limit on the number of items each person can bring or establish guidelines on the quality of the clothing items.

3. Arrange the space: Find a suitable space to hold the clothing swap. This could be someone's home, a community center, or a school classroom. Make sure there is enough space for people to try on clothes and sort through items.

4. Organize the clothing: On the day of the clothing swap, organize the clothing by type (e.g., shirts, pants, dresses) and size. This will make it easier for participants to find items they are interested in.

5. Swap away: Once everyone has arrived and the clothing is organized, the swapping can begin! Participants can take turns selecting items they want to swap for, or you can use a more structured system like a clothing swap game.

6. Donate leftover items: After the clothing swap is over, there may be leftover items that no one wants to take. Encourage participants to donate these items to a local charity or clothing donation center. By organizing a clothing swap, participants can have fun while reducing textile waste and promoting sustainable fashion practices.

7. Invite participants: Invite friends, classmates, or colleagues to participate in the clothing swap. You can use social media, email, or flyers to spread the word.

6. Reduce washing frequency

Reduce the frequency of washing clothing items by wearing them more than once or spot cleaning. This can reduce water usage and extend the life of clothing items. It just takes some of your time devoted to spot cleaning your clothes but it is well worth it.

Don't use your washing machine so much. Try spot cleaning your clothes by hand.

7. Air-dry clothes

Line dry clothes instead of using a dryer to save energy and reduce emissions. Doing so also helps with the long life of your clothes. If you have to use your dryer, then try using it once in a while.

Hang your clothes out to dry if you can instead of using your dryer.

8. Use ecofriendly laundry detergent

Choose laundry detergents made from natural or plant-based ingredients to reduce water pollution. Make sure to read the ingredients of your detergent to make sure that they are made from natural or plant based ingredients. Amazon carries different ones with great reviews. Some examples are Earth Breeze, Ecos, and BioKleen Free & Clear.

Use ecofriendly detergents like this one posted because it helps the environment.

9. Recycle old clothes

Donate old clothes to charities, textile recycling facilities, or sell them to secondhand stores. It takes research to find the nearest textile recycling facility from your home. It also takes research to find the nearest second hand store that will pay you for your donations. There aren't many around but there are some out there.

10. Educate others

Spread awareness about the importance of sustainable fashion through social media, blogs, or by organizing events in the community.

After you have read this book, you can educate others around you on fashion sustainability.

Glossary

Upcycling: Taking something no longer desired and putting new life into it.

Repurpose: The act of adapting something for a different purpose.

Sustainable Fashion: Fashion that is created ethically and eco friendly. Sustainable fashion brands use eco-friendly materials, such as organic cotton, hemp, and recycled fabrics, and employ production.

Textile recycling facilities: Places where streams of waste are received and turned into other recycled items.

Eco-friendly laundry detergent: Environmentally friendly detergent that is plant based and has non toxic ingredients.

Fast Fashion: Refers to the trend of producing low-cost, low-quality clothing quickly and in large quantities to meet the ever-changing demands of consumers. Fast fashion retailers prioritize profits over sustainability, which leads to a wasteful and harmful cycle of production and consumption. This leads to an increased amount of textile waste in landfills, as well as significant pollution of air, water, and soil due to toxic chemicals and greenhouse gas emissions.

Recycle: Converting things (waste) into something else that is used as new items.

Emissions: The production and discharge of something, especially gas or radiation.

Pollution: Something that is released into the environment that is harmful to living things.

Embroidery: The art of decorating fabric or other materials using a needle to stitch thread or yarn. It can incorporate other materials such as pearls, beads, quills, and sequins.

Patching: Mending something most likely clothing that has a hole in it by placing a piece of material over the hole and either sewing the material with a needle and thread or hot iron it in place.

Clothing swap: A clothing swap or CLOSWAP is a type of swap meet where participants exchange their used clothing for clothing of other participants that they will use. Clothing swaps are a great way to shake up your fashion style but also are considered an act of environmentalism.

It takes each of us to do our part when it comes to sustainable fashion. Play your part in your community because our planet needs us all to make sure that generations and generations will not be left with so much pollution. Today join DC in her practices of sustainable fashion. What will you do today to play your part?

DC is available for speaking engagements, workshops, alteration services, and sewing classes. Send her an email at Queenof3rs@gmail.com.

"Congratulations DC on completing your book. I found it interesting and I admire and appreciate your creativity. I learned some things about you. I think your history is profound. I learned some things about Revamping, Revitalizing and Reuse that I thought I knew already but did not. In fact you inspired ideas for me about clothing rentals. That idea had never crossed my mind. So thank you for your work and congratulations again."

--- Chris Jefferson

Save our Planet, Reclaim Revamp Repurpose: Ten Ways to Practice Fashion Sustainability is a book about one woman's efforts to open the world's eyes on ways communities around the world can practice fashion sustainability. This book contains the story of DC who has practiced sustainability since she was a little girl. And as the author of this book, DC gives tips and ideas on how you, the reader, can join in the efforts. Please join this great fight for our children and our grandchildren to live in a less polluted world. They deserve that and much more.

Made in the USA
Monee, IL
05 November 2026

67383546R00017